iScience Readers

The Human Body:
Train It Right

D0891209

by Emily Sohn and Katie Sharp

Chief Content Consultant
Edward Rock
Associate Executive Director, National Science Teachers Association

NORWOODHOUSE PRESS
Chicago, Illinois

Norwood House Press
PO Box 316598
Chicago, IL 60631

For information regarding Norwood House Press, please visit our website at
www.norwoodhousepress.com or call 866-565-2900.

Special thanks to: Amanda Jones, Amy Karasick, Alanna Mertens, Terrence Young, Jr.

Editors: Jessica McCulloch, Barbara Foster, and Diane Hinckley
Designer: Daniel M. Greene
Production Management: Victory Productions, Inc.

Paperback ISBN: 978-1-60357-307-8

The Library of Congress has cataloged the original hardcover edition with the following
call number: 2011011435

Printed in Heshan City, Guangdong, China.
190P—082011.

CONTENTS

Note to Caregivers:

Throughout this book, many questions are posed to the reader. Some are open-ended and ask what the reader thinks. Discuss these questions with your child and guide him or her in thinking through the possible answers and outcomes. There are also questions posed which have a specific answer. Encourage your child to read through the text to determine the correct answer. Most importantly, encourage answers grounded in reality while also allowing imaginations to soar. Information to help support you as you share the book with your child is provided in the back in the **Additional Notes** section.

Words that are **bolded** are defined in the glossary in the back of the book.

A Body at Work, a Body at Play, a Body at Rest

Most children like to run and play. Your body is like a machine. It gives you the energy to do many things. But like any machine, the body needs special care. In this book, you will learn about some **systems** in your body and how to keep them running fast and smooth.

Warming Up

Have you ever played or watched a soccer game? Soccer players do a lot of running, turning, and kicking. Trisha, Clay, and Juan have signed up to play soccer. Clay's little brother Ned wishes he could play, too. Each kid has a plan to get his or her body ready to play. Whose plan is best?

What do these boys need to do to make sure they can play their best soccer?

Trisha's Plan:

Rest. Trisha does not want to get too tired before the season starts. She plans to take lots of naps and play quietly.

Clay's Plan:

Eat. Clay wants to have plenty of energy for soccer. His plan is to focus on food. He will eat healthful foods. And he will eat a lot of them.

Juan's Plan:

Exercise. Juan wants to make his **muscles** strong. His plan is to move his body every day.

Ned's Plan:

Rest, Eat, and Exercise. Ned thinks it's good to run and kick the ball every day. He suggests sleeping a lot and eating healthful foods. The older children think their ideas are better.

Who has the best plan? Whose body will be best prepared to play when soccer season kicks off?

Materials
- cardboard
- scissors
- hole punch
- 2 balloons, one blue and one red
- tape
- 4 brads or fasteners

How Do Bones and Muscles Work Together?

You will make a model of an arm. It will show how muscles work together. Cardboard strips will represent the main arm bones. Balloons will be the main arm muscles.

cardboard strips taped together and punched with holes

Cut two strips of cardboard, each six inches (15.2 centimeters) long and two inches (5.08 centimeters) wide. Have an adult punch a hole in each strip. Put the hole about an inch and a half (3.81 centimeters) in from one end. Tape the strips together so the holes are three inches (7.6 centimeters) apart.

Think of one piece of cardboard as the upper arm. The other is the lower arm. The tape joins them at the elbow **joint.**

Have an adult punch a hole through both ends of each balloon. Attach one end of each balloon to opposite sides of the upper arm with brads or fasteners. Attach the other end of each balloon in the same way to the lower arm.

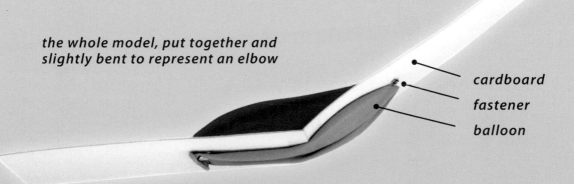

the whole model, put together and slightly bent to represent an elbow

cardboard

fastener

balloon

The red balloon on top is the **biceps** muscle. The blue balloon on bottom is the **triceps** muscle.

Bend the arm at the elbow. What happens to each balloon? Can you see how they work together to make the arm move?

Your skeleton holds you up. It also protects the organs inside your body. Your skull protects your brain, for example.

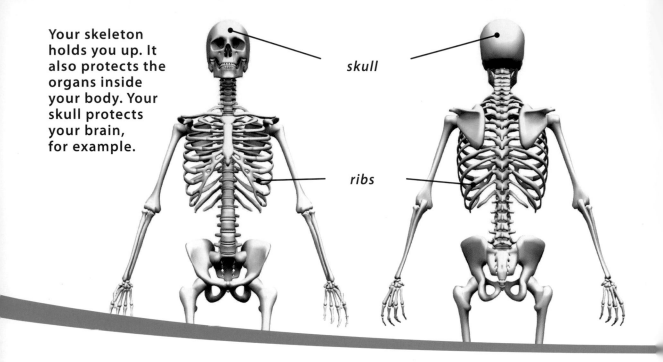

skull

ribs

There are 206 bones in your body. They make up your **skeleton.** Bones are pieces of rigid, or stiff, tissue.

Standing Tall

The skeleton supports the body and gives it shape. It is like the framework of a building. Without its frame, a building would collapse. Without a skeleton, you could not stand, walk, or sit.

Your skeleton also protects your **organs.** For example, in the chest area, the **ribs** protect the lungs and heart.

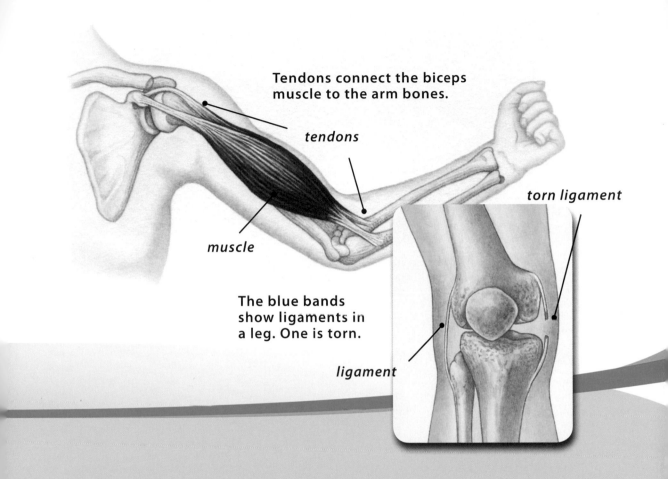

Tendons connect the biceps muscle to the arm bones.

tendons

muscle

torn ligament

The blue bands show ligaments in a leg. One is torn.

ligament

Bones can't move by themselves. They get help from muscles. Help also comes from strong straps of tissue called **ligaments.** These tissues connect bones to each other.

Tendons connect bones to muscles. Bones come together at joints. Hips and knees are joints that work hard in a soccer game.

Look at the arm you made in the Discover Activity. What do the brads or fasteners represent?

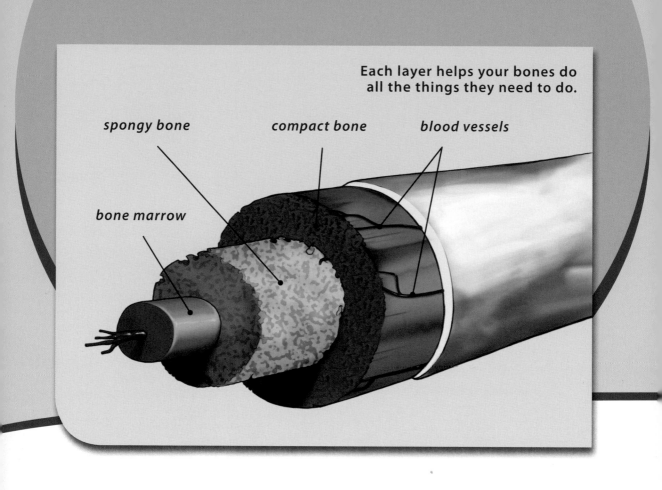

Each layer helps your bones do all the things they need to do.

spongy bone   compact bone   blood vessels

bone marrow

Inside Bones

Bones are more than meets the eye. First, there is an outer layer. It is full of **nerve cells** that carry messages between the bones and the brain. This layer also holds **blood vessels.** They carry blood to the bones. Blood is full of **nutrients.**

Underneath the outer layer is compact bone. This layer is thin. But it is hard and tough.

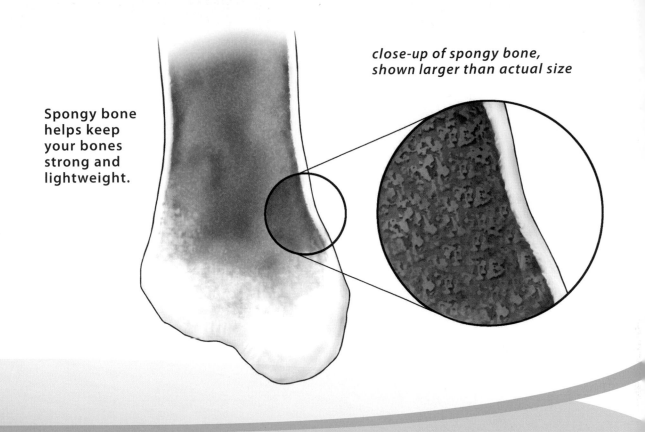

Spongy bone helps keep your bones strong and lightweight.

close-up of spongy bone, shown larger than actual size

Look deeper and you will find a layer of bone with little holes in it. It looks like a sponge. This is called spongy bone. The holes keep your bones light enough for you to move them.

In the middle of your bones, there is a layer that looks like jelly. This is **bone marrow.** It helps the body make blood.

Recipe for Strong Bones

To keep bones healthy and strong, you need to feed them well. They need **calcium.** This nutrient appears in milk, cheese, yogurt, and even ice cream. Broccoli, spinach, almonds, and other plant foods have calcium, too.

There's a lot of calcium in cheese. But there's a lot of fat in cheese, too, so don't eat too much!

When you eat these foods, your body breaks them down. This sends calcium and other nutrients into your blood. Blood travels to all parts of your body. Teeth and bones store calcium. The nutrient keeps them strong. Calcium is also good for your muscles, heart, and other body parts.

Think back to the iScience Puzzle. Which plans probably involve calcium?

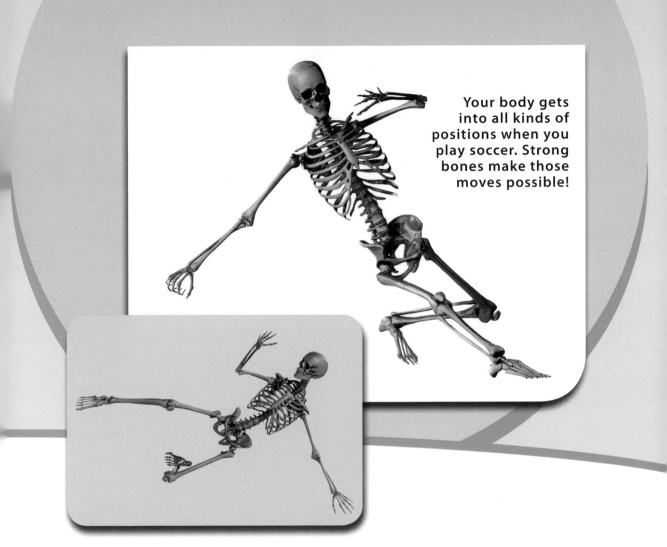

Your body gets into all kinds of positions when you play soccer. Strong bones make those moves possible!

Bone Power

Soccer players are always moving. Behind every kick, block, jump, and turn, bones are hard at work. Stronger bones make for stronger players.

Look at the model you made in the Discover Activity. You could represent weak bones with floppy cardboard. What would that do to their movement? How might weak bones affect the muscles?

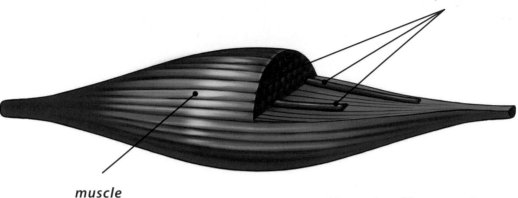

fibers that make up muscle tissue

muscle

Many tiny fibers work together in your muscles. That makes them stronger than if they were made of just one big fiber.

Bones are useless without muscles to move them. Muscles are made up of long strands called muscle fibers. Each fiber is thinner than a strand of hair. There are many bundles of fibers in a muscle. Tough tissue wraps them up.

Strength Control

Muscles are like players on a field. They belong to teams. And each has its own job. There are three kinds of muscle: smooth, cardiac, and skeletal.

Smooth muscle lines the walls of the stomach and intestines. These muscles **churn** food. They move the chewed-up mush along. You cannot control smooth muscle.

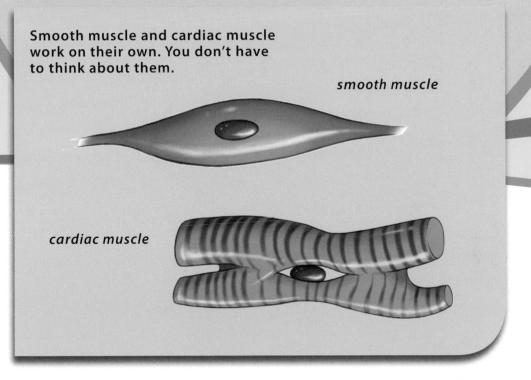

Smooth muscle and cardiac muscle work on their own. You don't have to think about them.

smooth muscle

cardiac muscle

Cardiac muscle is found only in the heart. It pumps blood through the body. You can't control cardiac muscle, either. The heart beats all by itself.

There are about 640 skeletal muscles in your body. You can control all of them. That's a lot of control!

skeletal muscle

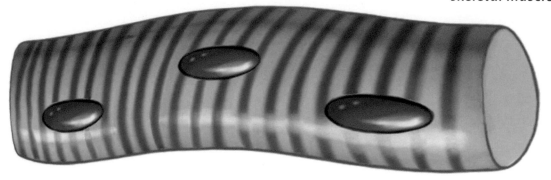

Skeletal muscles attach to bones. These muscles help you stand up, balance, and move. They also give the body its shape. You have skeletal muscles all over your body, and there are a lot of them. In fact, they make up about half your **weight**! You can make these muscles move when you want them to. That kind of control helps in a soccer game.

Think back to the iScience Puzzle. Which plans involve the skeletal muscles?

How the Joints Work

Joints make us move like people, not like robots. Joints occur where bone meets bone. They allow body parts to bend and twist. Elbows and hips are joints. What are some other parts that bend and twist?

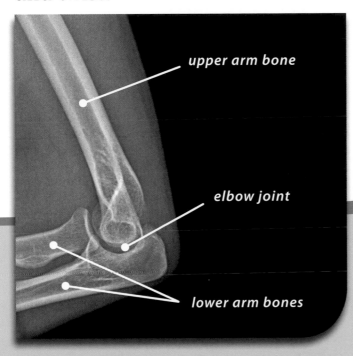

upper arm bone

elbow joint

lower arm bones

This X-ray shows the elbow joint, where the bone of the upper arm meets the bones of the lower arm.

Connected

Inside a joint, ligaments hold bones together. Ligaments are tough. They are also stretchy. They surround the ends of bones. And they bind bones together. Look back to page 11 to see a picture of ligaments.

How do you think ligaments help you play soccer?

Joint Jobs

Joints come in many forms. Three common types are hinge, pivot, and ball-and-socket joints. Hinge joints allow movement in just one direction — back and forth. Elbows and knees are hinge joints. Try bending your knee from side to side instead of forward and backward. You can't do it, can you?

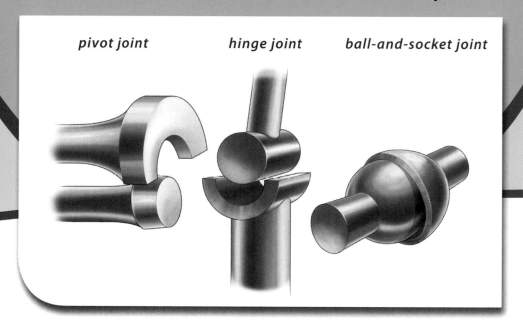

pivot joint   *hinge joint*   *ball-and-socket joint*

The neck is a pivot joint. It provides twisting powers. Ball-and-socket joints allow movement in many different directions. The hip is a ball-and-socket. So is the shoulder. Both can move in lots of ways. What if your hip had a hinge joint instead of a ball-and-socket joint?

How do your joints help you play soccer?

Connecting to History

Artificial Legs

Most people have two legs and two arms. Arms and legs are called limbs. But sometimes people are born without a limb. Other times, people lose a leg or an arm in an accident or during a war.

Long ago, wooden pegs were used to replace missing legs. The pegs were stiff. They helped people walk and balance. But they could not bend like knees and ankles do. They didn't feel the ground like feet and toes do. They couldn't tell hot from cold. They weren't much like real legs.

Artificial legs can do almost anything that real legs can. This man is using one to rock climb!

Today, **artificial** legs are a lot like real legs in the way they work. Some, but not all, look like real legs, too. To make them, doctors and engineers have studied the way bones, muscles, and joints work. These legs cannot feel pain or heat, like real legs do. But they can do some things that real legs can't. They may let people jump higher than before. And they don't break as easily as real legs do.

Scientists are now working to make body parts that can really feel things. They will be even more like our own fingers, feet, and toes.

How Muscles, Bones, and Joints Work Together

Muscles, bones, and joints work together to make the body move. Look at the arm you made in the Discover Activity. It models the bones and muscles of an elbow joint.

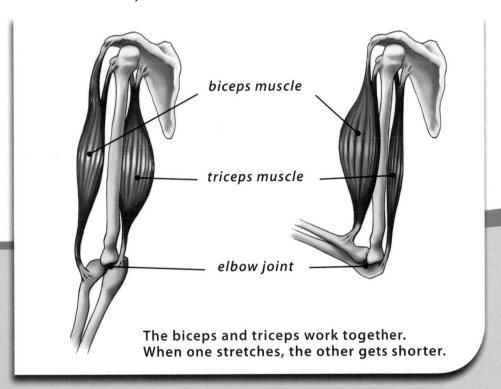

biceps muscle

triceps muscle

elbow joint

The biceps and triceps work together. When one stretches, the other gets shorter.

The muscles of the upper arm are the biceps and triceps. They work as a pair. When one gets longer, the other gets shorter. Muscles work by pulling. They don't push.

What would happen to the arm if you took away any one part of the model?

Your muscles work together to move your bones. That helps you throw the ball right where you want it to go.

Throw-in

You're playing soccer. The ball goes out of bounds. It's time to throw it back in. First, your brain sends a message to the arm muscles. It tells the biceps to **contract,** or get shorter. It tells the triceps to **relax,** or get longer. The bones move closer together. Your arms bend. As you throw the ball, the biceps relaxes. The triceps contracts. Your arms straighten. Play on!

Physical Therapist

At full speed, soccer players twist knees. Car accidents cause stiff necks. Some people have health problems that make it hard for them to even walk. Physical therapists help people who have problems moving. These experts know about special **exercises** that help bodies heal. One goal is to get rid of pain. Another is to prevent more injury. Physical therapists have to know about the human body to do their work. Why would that be?

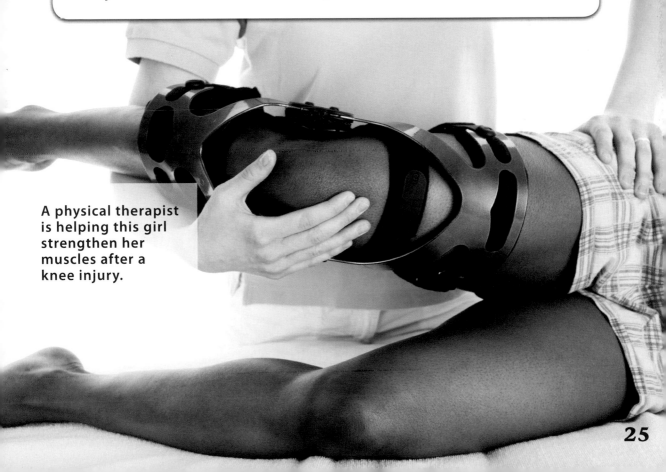

A physical therapist is helping this girl strengthen her muscles after a knee injury.

Healthful foods give your body what it needs to perform at its best.

Muscles and joints need nutrients, just like bones do. That's a good reason to eat well. Healthful foods include beans, pasta, beef, chicken, and fish. Fruits and vegetables are also good. So are low-fat dairy products. Nutrients in these foods help keep your whole body running well.

Muscles and joints also need exercise to stay strong. Movement makes muscle fibers grow. If you sit around all day, your joints get stiff. Your muscles get weak. Exercise causes tiny tears in the muscle fibers. That is part of what can make you sore. Waste products also build up in cells during exercise. Resting and sleeping help flush out the waste and repair the fibers.

Jogging helps build healthy, strong bones.

This girl is building muscle. Exercise can be fun!

Together, exercise and rest make muscles bigger and stronger. Your muscles may be sore when you begin exercising, but after a time the soreness should disappear.

Who Will Be Ready to Play?

Think back to the plans Trisha, Clay, Juan, and Ned came up with. They all want to be ready for soccer. Whose plan is best?

Trisha's Plan: Rest.

Clay's Plan: Eat.

Juan's Plan: Exercise.

Ned's Plan:

Rest, eat, and exercise.

Ned may be the youngest, but his plan is best. By moving every day, muscles, bones, and joints grow healthy and strong. Eating healthful foods gives the body all the energy and nutrients it needs. Rest helps the body recover after lots of hard, fun work.

You've learned that soccer players should exercise right, rest, and eat well to play their best. See what this plan can do for you. Pick an activity you like to do. It can be running, biking, or swimming. It can even be playing the piano or thumb wrestling! Now, grab a notebook or calendar and start a log.

Practice your activity every day for two weeks. Each day, test yourself. How fast did you go? Or, how many mistakes did you make?

Keep track of the foods you eat each day. Also, write down how much sleep you get each night. How do practice, rest, and nutrition affect your performance?

As you can see, it takes work and effort to be the best human body you can be!

Practicing every day is one good way for this boy to get his body ready to play basketball.

GLOSSARY

artificial: made by people.

biceps: the muscle of the front of the upper arm.

blood vessels: tubes that move the blood from one part of the body to another.

bone marrow: jellylike material inside some bones that plays a role in making blood.

calcium: a nutrient that helps build strong bones.

churn: to move around with force.

contract: become shorter in length.

exercises: activities to make the body strong.

joint: a place where two bones come together.

ligaments: strong straps of tissue that connect bone to bone.

muscles: tissues that produce movement in the body.

nerve cells: cells that carry messages back and forth between body parts and the brain.

nutrients: things in food that the body needs.

organs: certain body parts that do a certain function.

relax: lengthen, or become less tense.

ribs: the bones of the chest.

skeleton: the framework of the body made up of all the bones.

systems: groups of parts that are connected and work together as a whole.

tendons: tissues that connect muscle to bone and muscle to muscle.

triceps: the muscle of the back of the upper arm.

weight: the amount something weighs.

FURTHER READING

Human Body Q&A, by Richard Walker. DK Publishing, 2010.

Sensational Human Body Science Projects, by Ann Benbow and
    Colin Mably. Enslow, 2010.

How the Body Works, Nemours Foundation.
    http://kidshealth.org/kid/htbw/htbw_main_page.html

Your Gross & Cool Body, Discovery Kids.
    http://yucky.discovery.com/flash/body/

ADDITIONAL NOTES

*The page references below provide answers to questions asked
throughout the book. Questions whose answers will vary
are not addressed.*

Page 9: As the arm bends, the blue balloon stretches and the red balloon gets looser. When the arm unbends, the opposite happens.

Page 11: The brads or fasteners represent tendons.

Page 14: Clay's and Ned's plans involve calcium because they involve eating healthfully.

Page 15: The bones may not be able to hold up to the muscles pulling on them. They may break. Pulling on weak bones could make the muscles weaker, too.

Page 18: Juan's and Ned's plans involve the skeletal muscles.

Page 19: Your knees, ankles, shoulders, and neck bend or twist. Every time you move when playing soccer, your ligaments are working.

Page 20: If your hip were a hinge joint, you would be able to move your whole leg back and forth in only one direction. Your joints help you twist and bend so you can run after the ball and kick the ball.

Page 23: The elbow joint wouldn't work properly.

Page 25: Physical therapists need to know how the body works so that they do not hurt their patients by trying to make the joints and muscles move the wrong way or work too hard.

INDEX